BY VALERIE BODDEN

This edition copyright © Franklin Watts 2014
338 Euston Road
London NW1 3BH

Franklin Watts Australia
Level 17/207 Kent Street
Sydney NSW 2000

First published by Creative Education
P.O. Box 227, Mankato, Minnesota 56002
Creative Education is an imprint of The Creative Company
www.thecreativecompany.us
Copyright © 2010 Creative Education
International copyright reserved in all countries. No
part of this book may be reproduced in any form
without written permission from the publisher.

All rights reserved.

ISBN 978 1 4451 3034 7
Dewey number: 917.9'132

A CIP catalogue record for this book
is available from the British Library.

Printed in China

Franklin Watts is a division of
Hachette Children's Books,
an Hachette UK company.
www.hachette.co.uk

Design and production by The Design Lab
Art direction by Rita Marshall

Photographs by 123RF (Natalia Bratslavsky, Charles Shapiro),
Alamy (Inge Johnsson, Tom Till, Jim West), Corbis (Tom
Bean, Lester Lefkowitz), Dreamstime (Dallasphotography,
Dkarlsson, Jjmullen, Poco_bw, Tank_bmb, Tashka, Tsz01),
iStockphoto (Angelo Elefante, Eric Foltz, Mike Norton)

Every attempt has been made to clear copyright.
Should there be any inadvertent omission, please
contact the publisher for rectification.

# GREAT PLANET EARTH
# GRAND CANYON

## W
## FRANKLIN WATTS
### LONDON • SYDNEY

Colorado River

NORTH
AMERICA ★

A canyon is a long, narrow strip of low land that is surrounded by higher land. The Grand Canyon is found in the state of Arizona in the USA. The Colorado River flows through it.

The Colorado River rises in the state of Colorado and flows south-west towards the Grand Canyon.

The Grand Canyon is so big that it can be seen from outer space.

The Grand Canyon is one of the biggest canyons in the world. It is 446 km long. Most of the canyon is about 1.6 km deep. High rock cliffs surround the Grand Canyon.

The canyon's cliffs are made up of different layers of rock that have been worn away over time. Some of the layers are over two billion years old!

The Grand Canyon was created by the movement of the Colorado River. The water slowly wore down or **eroded** the rocky ground to form the canyon. Over millions of years, wind, rain and ice wore away the rocky cliffs to make the canyon wider.

Scientists estimate that the Colorado River began shaping the Grand Canyon around 17 million years ago.

The shape of the Grand
Canyon is always
changing, due to erosion.

The weather at the Grand Canyon's **rims** is warm in the summer and cold in the winter. Inside the canyon, the weather is always hot and dry. Some people live in villages at the bottom of the canyon, and there are many small towns near the canyon's rims.

In winter, there can be heavy snowfall at the rims of the Grand Canyon.

There are forests along the Grand Canyon's rims. Squirrels, mountain lions and black bears live there. Bald eagles (right) fly high above the canyon.

The forests along the Grand Canyon rims provide animals with food to eat, while birds such as Bald Eagles fish in the Colorado River.

About 50 different kinds of **mammals** live in the forests around the Canyon.

Few plants grow inside the dry, rocky Grand Canyon, but prickly **cacti** can be found there. Cottonwood trees grow along parts of the Colorado River, too.

Mountain goats and tortoises (above) eat the cacti that are found in the Grand Canyon.

**Native Americans** have lived in the Grand Canyon for around 4,000 years. They first lived in caves in the canyon's rocky cliffs. The first white people discovered the canyon around 500 years ago.

Early Native Americans dug homes in the rock that surrounds the canyon. They drew pictures in the rock, such as the one above.

Millions of people visit the Grand Canyon each year. Sometimes they damage the canyon and its surroundings. Some visitors leave rubbish on the ground. Nearby electricity power stations **pollute** the air. But today many people are working to protect the Grand Canyon.

Most visitors climb on the canyon's rocks without damaging or disturbing any of the plants or animals.

People visit the Grand Canyon to look at the canyon from its rims. Others may ride a **mule** down to the bottom. Some people go rafting on the Colorado River. Visitors are amazed by just how grand this canyon is!

Rafting on the river and riding mules are two good ways to see the Grand Canyon.

The Hoover **Dam** above the Grand Canyon slows the flow of the Colorado River.

**cacti** desert plants that have spines (sharp spikes) instead of leaves; a single plant is called a cactus

**dam** a wall built across a river to hold water back

**erode** to wear away the Earth's surface through the action of water, wind or ice

**mammals** animals that suckle their young

**mule** a small, sturdy animal that is a cross between a horse and a donkey

**Native Americans** people who lived in America thousands of years before European settlers arrived

**pollute** to make dirty with chemicals or other things that are bad for the earth, water or air

**rims** the high lands that are at the edge of the Grand Canyon

# Read More About It

*Looking At Countries: The USA* by Kathleen Pohl (Franklin Watts, 2011)

*Saving Wildlife: Desert Animals* by Sonya Newland (Franklin Watts, 2010)

# Index